LONGMAN BIOLOGY TOPICS

Searching for food

Y Deschet

Contents

Longman

Eating to survive

The instinct of self preservation makes all animals eat. Depending on the age of the individual and the species of animal, feeding requirements differ widely but no animal can escape this fundamental law of nature: it must eat to survive.

Firstly animals must eat to increase their own living matter, to grow. They also get from their food the energy needed for movement and for the working of their organs. Warm blooded[1] animals also need energy to maintain their body temperature.

All these overriding needs force every species into a never-ending search for food, either from plants or other animals.

Hunting scene in the African savannah. A buffalo has just been killed by the lioness. In the background hyenas await their share of the spoils, the remains of the big cats' meal. The latter, by killing off the weaker members of the species on which they feed, bring about a natural selection which benefits the hunted species.

[1] Warm blooded animals keep their bodies at a roughly constant temperature whatever the temperature of their surroundings.

Finding food

Whether herbivores or carnivores, all animals have a range of sense organs which keep them informed about various aspects of their immediate environment and, in particular, enable them to detect food.

Detection at a distance

In general, each species of animal has sense organs which are adapted to its environment and its way of life.

Thus, smell and hearing are particularly well developed in certain land-mammal hunters (for example, dogs) whilst sight is more important in others (for example, cats). The same senses are also very sharp in the usual prey of these hunters for whom survival often depends on the early detection of danger.

In birds, a group whose main characteristic is that they can fly, it is always sight which predominates over the other senses. Think of a bird of prey soaring hundreds of metres above the ground looking for its prey, and able to plunge on to it in a few seconds as soon as it is spotted.

On the other hand, visibility in water is much reduced. Even perfectly clear water is much less transparent than air, and the amount of light rapidly diminishes with depth so that below 300 metres there is complete darkness. The visual perception of aquatic animals is, therefore, always limited to a few metres or tens of metres. In contrast, the senses of smell and taste are well developed in fishes who are able to detect minute quantities of dissolved substances. There are many recorded instances of sharks converging from several kilometres away, towards a wounded fish or marine mammal which is bleeding in the sea. Experiments have shown that

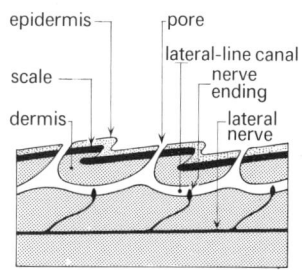

closing off the nostrils of these fishes renders them incapable of locating their prey. These nostrils have nothing to do with breathing; they are just tubes or holes lined with olfactory sense cells.

The lateral line can be seen clearly on this carp.

Fishes' sixth sense

Greatly enlarged cross-section through the lateral line.

Fishes also have what is sometimes called a 'sixth sense' because we have no equivalent sense: the lateral line. This line, often clearly visible on the sides of fishes, is made up of a longitudinal row of perforated scales. Each hole connects with a canal lined with nerve endings which detect changes in pressure and water movement. Anglers know that the slightest vibrations (like those caused by a voice) spoil their chances of making a catch.

5

Short range detection

Food which is close at hand is usually detected by the senses of taste and touch. Many animals, vertebrates and invertebrates, have organs to provide one or other of these senses (or both together)[1].

Many invertebrates with poor sight or none at all – and particularly fixed animals – rely entirely on taste and touch to tell them about their environment. The sea anemone can only 'feel' through its tentacles which are withdrawn into the body cavity as soon as its surroundings become unfavourable. Similarly, the mussel learns of changes in its environment through the sensitive edge of its mantle which can be seen in the gap between its shells. Any foreign body touching this mantle causes the shell to close immediately.

The 'sonar' of dolphins and bats

Dolphins and whales, which are mammals that have returned to the sea, have poor sight and sense of smell. However, their hearing is excellent, better

The eye of a cuttle-fish thought to be as effective as the eyes of vertebrates.

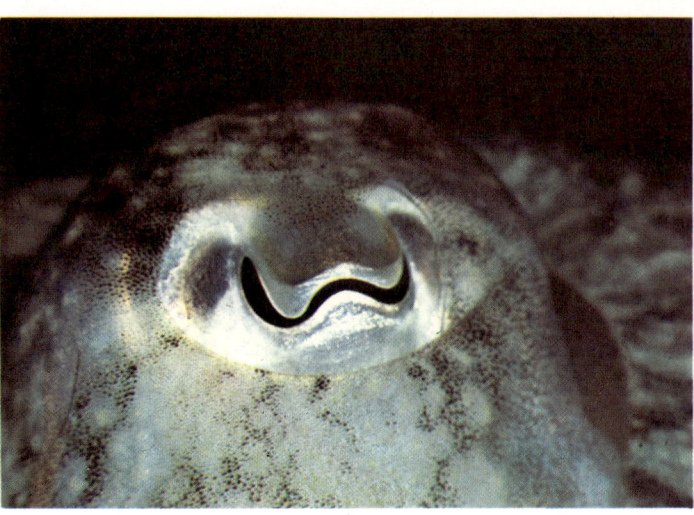

[1] The whiskers of cats, barbs of fish, feelers and antennae of insects, tentacles of molluscs and the tube feet of sea urchins.

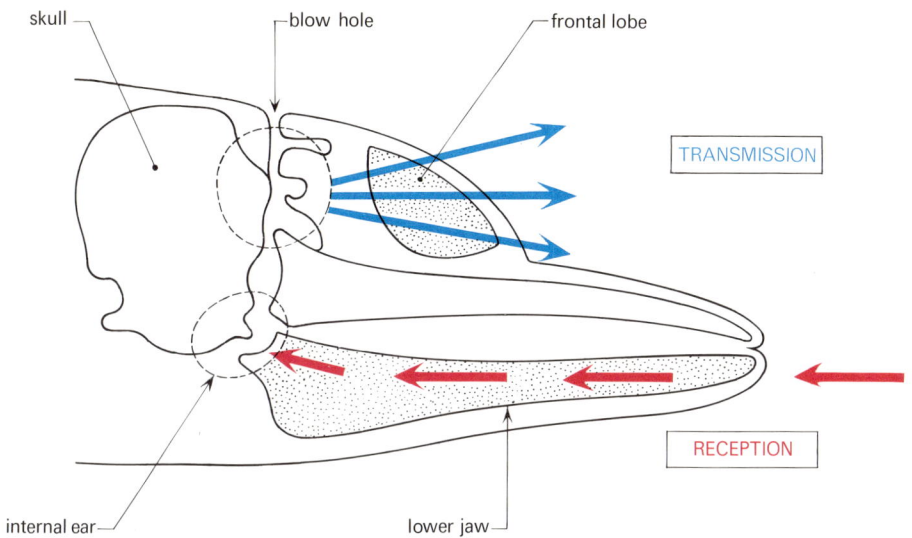

skull — blow hole — frontal lobe

TRANSMISSION

RECEPTION

internal ear — lower jaw —

than that of any other animal species. The dolphin makes a variety of sounds: clicks, whistles and cries inaudible to man (ultra-sounds). They are produced in the internal folds of the single nose hole, the blow hole. When they strike an obstacle, these ultra-sounds are reflected and their echo returns to the animal who made them. The nearer the obstacle (for example, a meal in the form of a fish), the more quickly and loudly the echo is sensed[1].

Like dolphins, bats make ultra-sounds which enable them to fly around obstacles by echo-location and to hunt insects at twilight in spite of their poor eyesight. This ability is shown by the following experiment. A bat is shut in a dark room with wires stretched between the walls so that the animal can just pass between them. If the bat is blindfolded it never flies into the wires but if its ears are blocked up it frequently knocks into the obstacles. Thus it is clear that the ultra-sounds given out by the bat are sensed by its ears, which are often very large.

A dolphin's sonar. The ultra sounds emitted from the area of the internal folds in the nasal orifice are amplified by the frontal lobe. Reception is through the lower jaw to the internal ear.

[1] The principle of echo location has been copied by man in the sonar equipment used in ships to explore underwater relief and to find submarines and shoals of fish. However, man's sonar lacks the precision of that of a dolphin which can distinguish between two spheres whose diameters differ by only a few centimetres or between two fish of the same size but different species.

The guinea pig moves its lower jaw constantly, using its incisors like wood chisels to reduce the hardest plant materials to small pieces which are then ground up by the molars.

Equipment for eating plants

Some animal species eat only plants; they are called herbivores. The way of life, teeth and digestive systems of such animals are well adapted to their feeding needs. The cellulose, and sometimes the silica, content of plants makes them fairly tough. Specialized tools, teeth or mouth parts, adapted for the purpose are needed to tear or cut and then grind plant food. In addition, the food value of plants is less than that of foods of animal origin, so that herbivores need to eat much more than carnivores. This imposes on them a continuous search for food. Finally, the difficulty of digesting vegetable matter means that plant eaters must have longer and more complex digestive systems than meat eaters.

Rodents

These are a group of mammals who are well adapted to feeding on hard plant materials (especially roots, seed and bark) although they are not always strict vegetarians. The characteristic feature of their dentition[1] is their long, curved incisors which are worn to a sharp cutting edge by the incessant movements of their jaws. The molars of rodents, capped with transverse ridges of enamel, look like a file. The constant wear on these teeth is counteracted by their continual growth.

Ungulates[2]

These animals use a range of grazing techniques depending on the pattern of their dentition. For example, a cow tears off grass by gripping it between the incisors of its lower jaw and the horny gum of the upper jaw, using its rough tongue to help. A horse, on the other hand, pinches tufts of grass between its upper and lower incisors, then pulls and cuts. However, ruminants like cattle, sheep, deer or giraffes and non-ruminants such as horses, zebras, rhinoceroses or elephants all have very similar molars. Their progressive wear produces a wear platform which is characteristic of the animal's age.

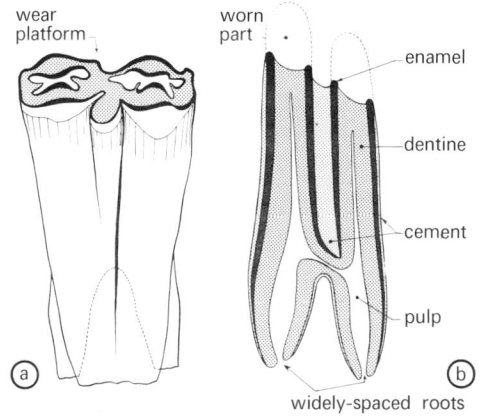

wear platform — worn part

enamel

dentine

cement

pulp

a   b

widely-spaced roots

A cow's molar. a. overall view; b. cross-section. The enamel ridges are at right angles to the main direction of jaw movement which makes them effective in chewing plants. Canine teeth are absent or much reduced in cattle as in rodents. A wide gap separates incisors and molars.

[1] Dentition is a word used to describe the number, types and arrangement of an animal's teeth.
[2] The ungulates are a large group of herbivorous, hoofed mammals.

9

Gerenuk. This long-necked antelope is particularly fond of the spiny leaves of acacias . . .

In the wild, ungulates usually live in herds and spend most of their time grazing, thus consuming an enormous quantity of vegetation. Accordingly, it might be thought that different species living in the same area would compete in this full-time search for food. But in general this does not happen because each has its own preferred foods. In African grasslands, zebras, gnus and other antelopes, gazelles, giraffes and elephants are often found side by side. All these species use the food resources of their environment in complementary ways. Some (giraffes and elephants) eat mainly leaves whereas others prefer different types of grass: tall for zebras, medium for gnus and very low for gazelles. Moreover, these herds are always on the move allowing the vegetation to grow again after their passage.

However, the clearance of immense areas for farming in recent years imposes considerable restrictions on the freedom of movement of such wild herds gathered in game reserves or national parks. In the case of elephants, once threatened with extinction from over-hunting, this restriction has created a new danger. Having few natural enemies apart from man, the elephant herds are once more thriving. No longer able to make their long migrations, they strip, break and tear up trees to feed on the leaves and young branches, leaving the vegetation no time to grow again. Thus they are unwittingly destroying their last refuges and the balance of nature in the reserves.

Digesting grass

The digestion of grass (and other vegetation) is long and difficult. All herbivorous mammals have very long digestive canals: 20 metres in horses and as much as 40 metres in cattle! Also cattle, like all ruminants, have a very special form of digestion. During grazing grass is swallowed without being

... whereas, the gnu prefers grasses of medium height.

11

chewed and is then stored in the rumen, which has a capacity of up to 150 litres. When the animal is resting small balls of grass are formed by another part of the stomach called the reticulum. These balls are returned, through the oesophagus, to the mouth. There they are chewed for a long time by the molars: the animal is ruminating. The grass travels backwards and forwards between the mouth and the rumen until it is completely ground up. Reduced to a pulp, the food enters the manyplies then the abomasum where digestion properly starts.

Diagram of a cow's stomach. The dotted line shows the path of unchewed grass; the solid line is the path of ruminated food.

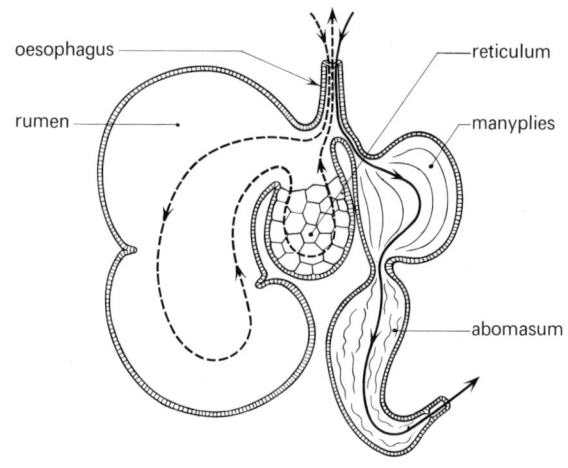

oesophagus

reticulum

rumen

manyplies

abomasum

It is believed that wild ungulates which ruminate have an advantage over those that do not. Ruminants spend less time grazing as the grass is swallowed without chewing. During this time a herbivore, its head down, busy chewing, is at its most vulnerable. Relaxing its habitual wariness, it can be surprised by an attack from a predator. It is easy to see the advantage of reducing this dangerous period. During rumination, the animal can chew at its leisure whilst keeping all its senses alert. In addition, this process can be interrupted at will, for example at the approach of danger, and continued later.

Invertebrates

Vegetarian invertebrates also have highly adapted 'equipment' for feeding. Snails and insects are good examples.

Look at the mouth of a snail which is 'nibbling' a lettuce leaf. At the top you will see a horny jaw cutting out small pieces of leaf. Put your ear close and you will hear rasping noises caused by the back and forth movements of a rough tongue, the radula, in the snail's mouth. Examination under a microscope reveals that the radula is covered with innumerable 'teeth' which work in a similar way to the molars of herbivorous mammals.

The mouthparts of insects are adapted to a wide variety of diets which may be liquid or solid, animal or vegetable. If you dissect the mouth of a grazing insect, like a grasshopper or a cockchafer, you will find strong chewing parts with characteristics similar to the teeth of herbivorous mammals. Insects like butterflies or bees have mouthparts which correspond to those of the grasshopper, but in contrast they are very different in shape. They are adapted for sucking up a liquid diet of nectar from flowers.

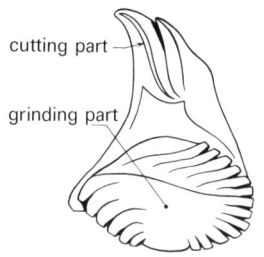

Front view of one of the pair of mandibles (jaws) of a cockchafer.

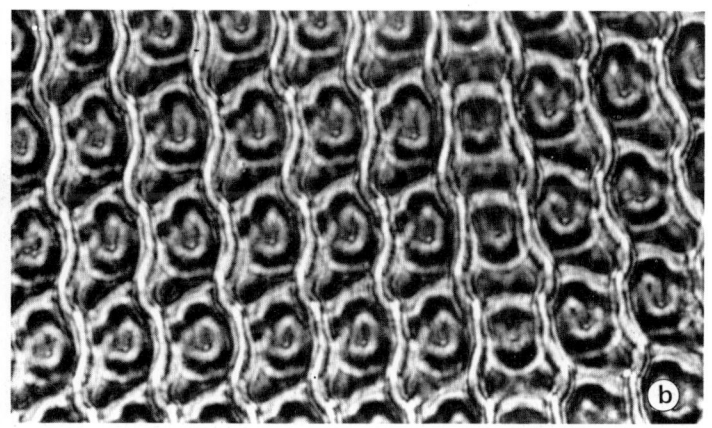

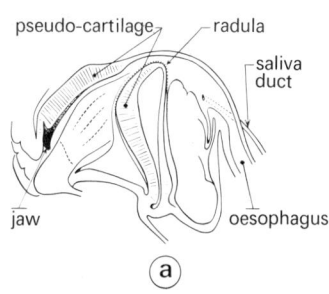

The chewing parts of a snail. a. vertical cross section of the mouth; b. magnified picture of part of the radula (× 300).

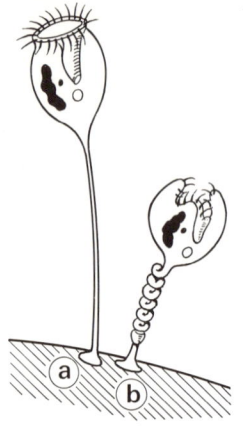

Predators' weapons and techniques

All animals, regardless of their size, which feed on other animals are called predators. In general, they are hunters well equipped to detect, capture and then eat their prey. But some are quite incapable of moving about. How can such animals feed themselves?

Sedentary predators

A whole range of different groups of animals include sedentary (fixed in one position) predators. In the microscopic world of protozoa there are two fixed giants: *Vorticella* and *Stentor*. Both have a crown of fine hairs (cilia) which wave incessantly, making small currents in the water to carry their food (microscopic plants and animals) to them.

In a similar way, many fixed invertebrates of the sea shore create water currents to enable them to extract from sea water the plankton on which they feed. These currents are set up by the vibrating cilia of the gills of lamellibrancs[1], the feathery gills of bristle worms or the modified legs (cirri) of acorn barnacles.

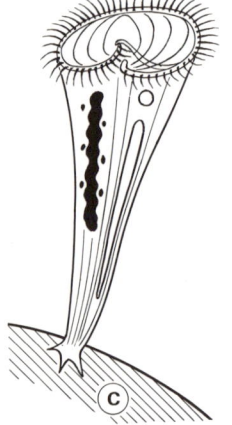

Fixed protozoan ciliates: *Vorticella* a. extended; b. contracted, and c. *Stentor*.

Free-swimming crustaceans in the larval stage, adult acorn barnacles remain fixed on exposed rocks, leading a life dictated by the rhythm of the tides. As the tide goes out, they withdraw inside their hard protective shells. When the sea covers them again, the mobile plates at the top of the shell open to expose a plume of cirri whose beating draws in a food-bearing current.

[1] Lamellibranchs are molluscs with two shells, like cockles, mussels and scallops.

The lion's teeth are very well adapted to a strictly carnivorous diet with large canines (fangs) and sharp, cutting molars. The jaws open wide and the powerful chewing muscles close them tightly on the animal's prey.

Hunting methods

Mobile predators use many different hunting techniques. The following three examples illustrate the most common behaviour patterns.

Cape hunting dogs, wild dogs of the African plains, hunt in packs for animals which are much larger than themselves, mainly gnus and zebras. They do not try to outrun their prey, but exhaust them by a long chase. Eventually a weaker member of the hunted herd tires and lags behind to be caught and eaten alive by the pack of hunting dogs. The cheetah uses quite a different technique. Usually hunting alone, it is a sprinter, capable of speeds of 100 to 120 kilometres per hour, but only over distances of a few hundred metres. Like the domestic cat, it must use a stealthy approach to reduce its distance from the selected prey (usually gazelles) to a minimum before launching an attack. If the gazelle does not run off in time, the cheetah will quickly catch it, roll it over in full flight and bite through the throat with its fangs.

In the sea, the cuttle-fish can take on a stripy appearance so that it blends perfectly with the seaweed in which it hides to ambush its victim. When some unsuspecting prey comes close enough, the cuttle-fish seizes it with its tentacles which are covered with suckers. The prey is carried to the cuttle-fish's mouth and torn up by its horny jaws which are shaped like a parrot's beak.

Thus the Cape hunting dog's stamina, the cheetah's high-speed bursts and the cuttle-fish's ability to camouflage itself are used in three quite different hunting techniques: pursuit, stalking and ambush.

A kestrel with its prey

Predators' natural weapons

Predators are always remarkably well equipped to deal with the prey on which they feed.

Birds of prey, whether they hunt by day or by night, are the principal predators of small rodents. In addition to their excellent eyesight, these birds have two other common characteristics: a powerful hooked beak, and curved talons which are strong and sharp, making them particularly good for capturing and killing the prey.

Comparing insect eaters which are as different as the anteater and the chameleon, we notice that both have long, sticky tongues which they throw out

The giant anteater, a mammal of the South American plains and forests, feeds on ants and termites. These are dislodged from their hiding places by the anteater's claws and snapped up by its long, sticky tongue which, like the chameleon, it can project at great speed.

considerable distances to catch their favourite food.

In the first example (birds of prey), all the species of a group of animals show the same characteristic adaptations. On the other hand, in the second example (insect eaters) there is a similarity of adaptations to the same diet in two unrelated animals (a mammal and a reptile). But, similar problems are not always tackled in the same way. The common whelk and the starfish illustrate that there are sometimes several solutions to the same problem. Both these invertebrates cause considerable damage

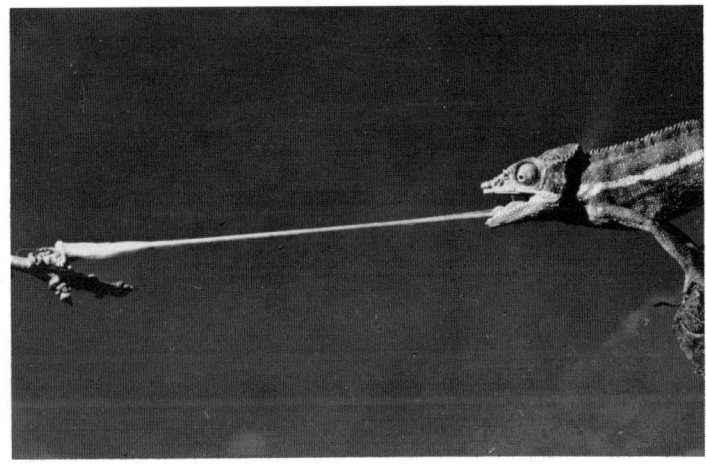

A chameleon capturing an insect.

in oyster and mussel beds. Their victims are immobile but well protected by hard shells. The predators do not need to move fast but must have a means of overcoming their victim's natural protection. The whelk has a specialized tube which it extends from its mouth to pierce the shells. The starfish opens the mollusc shells by enveloping them in its arms and pulling steadily with the suckers at the ends of its tube feet. After a few minutes, the shells are forced apart and the starfish's stomach comes out of its mouth to slip between the partly open halves of the shell. Digestion then takes place outside the starfish, the digested food being absorbed inside the shellfish by the stomach wall.

The forelegs of the praying mantis appear to be in an attitude of permanent prayer. In fact, they are perfectly adapted to seizing and immobilizing other insects which the mantis eats alive.

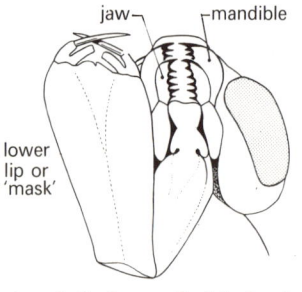

The head of a dragonfly (*Aeshna*) larva with its mask partially unfolded.

Carnivorous insects have a whole arsenal of weapons which, although miniature, are fully effective for the size of prey hunted. Let us look at just two highly developed capture mechanisms: the mask of an *Aeshna* larva and the grasping forelegs of the praying mantis.

Aeshna is a large dragonfly which is common around streams and ponds. Its aquatic larva is an efficient ambusher which captures its prey with the help of a modified lower lip folded in front of its mouth like a mask. There are two gripping claws at the end of this weapon which is rapidly extended when a victim approaches.

Traps and tools

Very few animal species make use of anything which might be called a 'tool'. Those tools that are used are very primitive being no more than simple stones or sticks. Nevertheless, the behaviour of the animals using them is interesting.

The Egyptian vulture, a small vulture of the African savannah, feeds on ostrich eggs which it breaks by lifting stones in its beak and dropping them on the shell. The sea otter uses a similar technique to break open the shellfish, crustaceans and sea urchins on which it feeds. It returns to the surface after every expedition carrying a large pebble from the sea bed as well as its catch. Lying on its back, the sea otter uses its tool like an anvil or a hammer, enjoying a sea-food meal deprived of its natural protection.

The traps devised by some arthropods[1] are even more elaborate. Many spiders, such as the common garden diadem spider, patiently spin webs designed to capture flying insects. When a victim flies into the trap, it becomes entwined in the sticky threads, and its struggles to escape warn the spider of its presence.

A spider's web. An unwary insect has been caught in the trap and immobilized in a cocoon of sticky silk.

[1] Spiders, insects and crabs are all arthropods.

Ant lion fly larva.

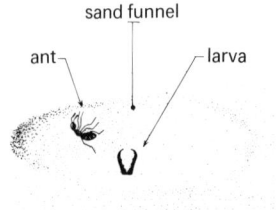

The trap built by the ant lion fly larva.

The spider rushes up and immobilizes its prey by covering it with sticky silk threads and then paralyses it by an injection of poison. The bound and paralysed insect is then eaten. The spider pierces the hard body with its jaws and through the wound it injects digestive juices which liquify the flesh. The digested products are then swallowed. This external digestion may remind you of the earlier examples of the whelk and the starfish.

A quite different type of trap is dug in fine sand by the larva of the ant lion fly. It makes a shallow funnel, a few centimetres across, and hides in the bottom. As soon as an insect slips down the sandy sides the larva seizes its prey and paralyses it with its claws then digests it in the same way as the diadem spider.

Poison and electric shocks

Poisoned weapons are widely distributed throughout the animal kingdom. Bee and wasp stings are well known, that of scorpions is rightly feared and we have just described two other arthropods (diadem spider and ant lion fly) which use poison to paralyse their prey.

Coelenterates (another important group of invertebrates) are all equipped with stinging cells (nematocysts). These cells cover the tentacles of fresh-water *Hydra*, sea anemones and the colonial polyps which form coral. They are also the source of the very unpleasant stings which can result from touching jellyfish. When a prey brushes against the extended tentacles of a sea anemone, it triggers the firing of these miniature harpoons which inject a paralysing liquid into the skin of the victim. Retraction of the tentacles then brings the prey level with the anemone's mouth where it is swallowed.

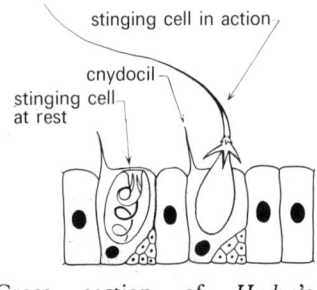

Cross section of *Hydra*'s epidermis.

The poisoning equipment of adders is made up of two fangs pierced by a duct through which flows poisonous saliva, produced in the poison glands (modified saliva glands). Normally the fangs rest against the palate but when the snake opens its mouth to bite, they swing forwards and, at the same time, a muscular contraction squeezes the poison gland like a rubber bulb, injecting poison into the wound.

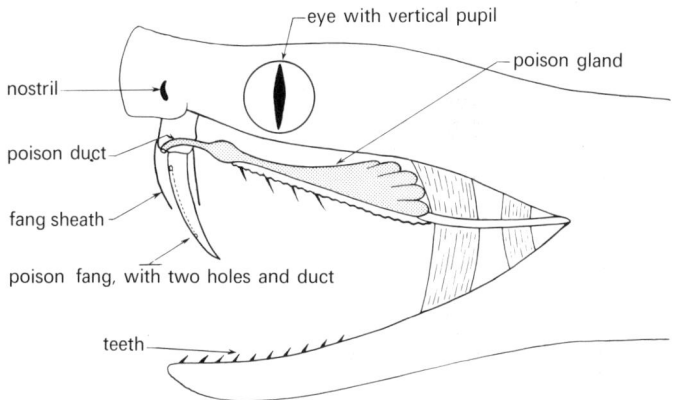

The poison system of an adder.

Unlike the use of poison, the ability to give electric shocks is extremely rare amongst animals. Electric rays have, on either side of their head, two electric organs (modified muscle fibres) capable of producing a discharge of several hundred volts which is used to stun victims. Beware, the swimmer foolish enough to tread on one!

To eat and not be eaten

. . . that is the rule in nature. Animals, therefore, have a certain number of defence mechanisms to escape from their predators. We can distinguish four basic types: passive defence, camouflage, flight and . . . attack.

Passive defence

This consists of making yourself incapable of attack within a discouraging natural protection. Gastropod and lamellibranch molluscs seem well protected within their hard shells but we have seen that, in practice, some of their predators have developed the ability to pierce (whelk) open (starfish), or smash (sea otter) their shells. Crustaceans do not have a permanent armour covering as they must leave this when they moult in order to grow. This is always a very dangerous time for these animals and even aggressive crabs hide under rocks for a few days while their new shell hardens. The hermit crab seems to have found it more advantageous to protect its soft abdomen in an empty gastropod shell. But he, too, must move home as he outgrows its protection.

The spiny protection of the sea urchin is permanent whereas the hedgehog must roll up into a ball at the first sign of danger to take advantage of its spiny covering. Similarly, the porcupine fish faced with danger, inflates itself with water and bristles with spines like the mammal from which it gets its name.

The spines of the hedgehog provide an effective passive defence against a large number of its enemies ... except the motorcar which, every year, squashes thousands of these likeable and useful insect eaters.

The Alpine hare lives in the Alps above 1200 m. Its fur which is brown in summer, turns white in winter, providing excellent adaptation to the changes in the hare's environment.

Camouflage

To merge with their environment, some animals adopt its colouring. The fur of ermines and Arctic hares changes colour to match seasonal changes in the environment; flat fish can imitate the sandy or stony appearance of the sea bed on which they live.

Other species copy not only the colour but also the shape of natural objects on which they usually live. This imitation may be so good that they become practically invisible when they are not moving. For example, stick-insects and leaf-insects, as you might guess from their names, look just like twigs and leaves.

The bright colourings of fish which live around coral reefs and the contrasting stripes of the zebra's coat are a more unexpected form of camouflage. In fact, these designs break up the animal's outline in

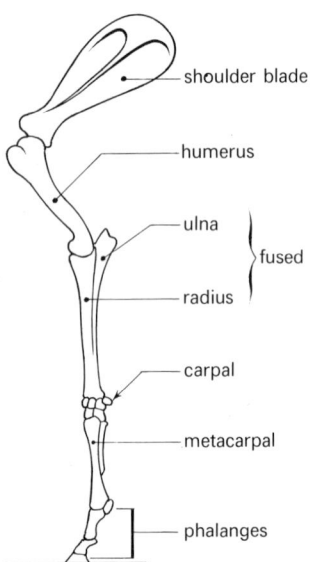

Skeleton of the front leg of a hoofed animal (ungulate).

- shoulder blade
- humerus
- ulna }
- radius } fused
- carpal
- metacarpal
- phalanges

Wild ungulates are highly adapted for running in a number of ways which can be seen clearly in zebras: a small, muscular body; legs which are lengthened by the special bone arrangement shown above and improved further by a reduced number of toes; powerful muscles for moving their legs.

the vision of a possible predator. This confusion is even more effective when the animals live in groups: within the shoal of fish or the herd of zebras, each individual merges into a vague, moving mass.

Flight

This is certainly the commonest and one of the most effective solutions to the problem of self-defence. A large number of species have acquired great skill at moving themselves quickly allowing them to outpace their enemy, to regain their lair or to find safety in a temporary shelter.

In air or in water a streamlined shape is a basic requirement for high speed movement. The machines which we use to overcome our own inability to fly or swim, in effect, copy those species which are best adapted to rapid movement in these media.

Similarly, the principle of jet propulsion is not a human invention. The scallop leaps away when threatened by forcing the two halves of its shell together. A squid moves quickly backwards by forcing a jet of water out of its breathing cavity; at the same time a dark cloud is ejected by the ink gland to hide the retreat from a predator.

Flight does not seem possible once the hunter has seized its prey. In fact, some animals can still escape by voluntarily abandoning part of themselves: the lizard's tail, the crab's legs and the arms of starfish. This 'self-amputation' is accompanied by the ability to regrow the abandoned part.

Attack

'Attack is the best form of defence.' In nature, this saying is very seldom true. Even if the prey has any weapons (horns, claws, talons, hooves, etc.) these will usually be less effective than those of the predator. Their main use will be in fights with members of the same species.

Finally, the most subtle defence strategy is surely that of making an attacker believe your strength to be greater than it is so that you do not have to use it at all. Indeed, before they start fighting, many animals try to frighten their adversary by aggressive attitudes or noisy displays. Fur which stands up, puffed out feathers or shrill war cries are sometimes enough to deter an attacker.

This lizard has lost its tail and regenerated an abnormal replacement.

Living together to eat

So far we have looked at the special abilities developed by different species in their search for food. Some have highly developed sense organs, some have powerful weapons, and others have effective means of defence. It can also be a great advantage for two species to live together (associate) so that their individual abilities can be used jointly for their mutual benefit.

Commensalism

In this type of association only one of the partners really benefits. For example, there is a small, red crab which shelters and feeds inside mussel shells. The mussel does not appear to suffer from the presence of this stranger, but it does not gain any advantage either.

The association between sharks and pilot fish benefits only the latter. Swimming close to the shark, they are carried along in the slipstream of their powerful team-mate and they feed on the remnants of the predator's meals.

Mutualism

Both partners benefit from living side by side in this type of voluntary association. A well known example is the clown fish and its anemone. This small fish of coral reefs hides among the tentacles of the anemone and is probably protected against, or immune to, the poison from its host's stinging cells.

The hermit crab and the sea anemone are an example of mutualism which often become true symbiosis: the hermit crab carrying around its inseparable anemone.

In return, the bright colours of the fish attract victims to its partner with which it shares the meal.

Mutual aid extends to a group of animals within the mixed herds of zebras and gnus which are sometimes joined by other antelopes and ostriches. We have already seen that living in groups is itself a source of safety. An isolated individual is always more vulnerable than a herd which can detect danger no matter from what direction it comes. In a mixed herd the safety is even greater as each species uses its special skills. In open country, the height and excellent eyesight of the ostriches enable them to keep watch over a vast territory; in scrub, the denser vegetation limits the field of vision of the birds and these then benefit from the sense of smell of the antelopes and the hearing of the zebras.

Sacculina, a parasite of crabs, is a good example of degeneration, an important adaptation of parasites. It is little more than a sac fixed to the crab's abdomen with many fine tubes penetrating the host's body from which it extracts its food. Sense organs, digestive apparatus and means of movement have completely disappeared. *Sacculina* retains only well developed reproductive organs which are necessary for the survival of the species.

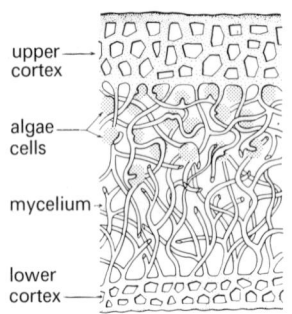

upper cortex

algae cells

mycelium

lower cortex

Cross-section of a lichen.

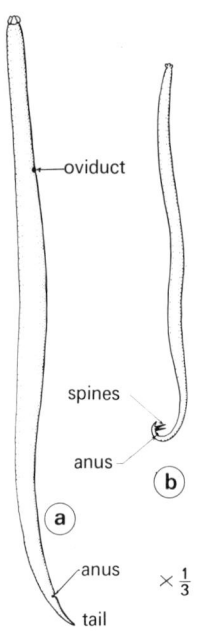

oviduct

spines

anus

a

b

anus

×⅓

tail

Ascaris (round worm)
a. female;
b. male.

Symbiosis

This type of association also benefits both partners but it is obligatory, neither being able to live without the other.

Some hermit crabs carry a sea anemone on the shell which they have chosen for a home. In this way the crab increases the protection of its soft abdomen. The sea anemone benefits by being carried around and by feeding on the remnants of the crab's meals. Although it is usually no more than simple mutualism, this association becomes symbiotic for some species of hermit crab and sea anemone which are never seen separately. Also, when the hermit crab outgrows its home and has to move, its first job is to carry its companion to the newly chosen home.

Lichens are plants made up by the association of a green alga and a fungus. Like all plants containing chlorophyll, the alga uses sunlight to make organic food which the fungus cannot make for itself. In return, the fungus provides the alga with a protective covering and the moisture which it needs. This association is so successful that lichens can live in places where no other plants could survive. Colonizing bare rocks, they can resist an arctic climate as well as a desert one. They are truly pioneering plants.

Parasitism

Only one partner benefits from parasitism. It lives entirely at the expense of the other and in extreme cases kills it.

The parasite may be external; such as ticks, fleas, mites, greenfly, leeches and lampreys which suck blood, or sap. In other cases the parasite may spend the whole or part of its life cycle inside its host.

Examples of this behaviour are gall wasps which cause oak galls and the worms which live in the digestive canal of mammals (tapeworms, round-worms, etc.).

Myxomatosis
or the interdependence of living things

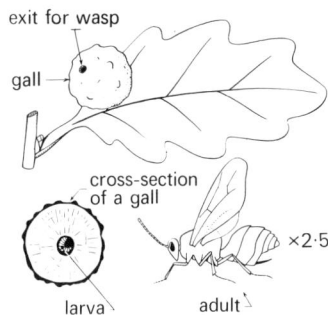

The oak gall-wasp and its gall.

Once upon a time there was a beautiful area of countryside which was a real hunters' paradise. Rabbits flourished there on the abundant grass, roots and bark. However, farmers were not at all keen on their invasion of farms and gardens or on the damage they caused to young trees. So one day they decided to get rid of some of these prolific rodents by introducing a rabbit disease, myxomatosis; little did they expect the rapid and drastic results which followed! Within a year, 99 per cent of the rabbit population had been killed off. Harvests improved and trees began to grow again ... but unfortunately other effects of this slaughter soon began to appear. It was impossible to control the myxomatosis epidemic which rapidly spread to wild rabbits throughout the land, even tame animals were affected. Foxes, the rabbits' main predators, soon had to look for other sources of food ... hen-houses were plundered more and more frequently ... in turn small birds disappeared ... allowing insects to multiply. Disaster for fruit trees! Buzzards also prey on rabbits, and they became short of food and produced fewer chicks. Their numbers declined rapidly. The farmers, who thought they were protecting their gardens and woodlands saw, instead, the devastation of their orchards, hen-houses and rabbit hutches. And it was many years before the hunters could go rabbit shooting again.

This is a true story which began in Solange, a district in the west of France in 1952. By the end of 1953 the disease had spread throughout Europe, including Britain where there was similar de-struction. Think about the lessons of this story.

Firstly, it illustrates several examples of what biologists call food chains, which are set out thus:

Root → Rabbit → Fox

or:

Seed → Fieldmouse → Buzzard, etc.

Each arrow means 'is eaten by'.

These food chains show clearly how the different species of a habitat are united (like the links of a chain) by their need for food. We say that these living things are interdependent.

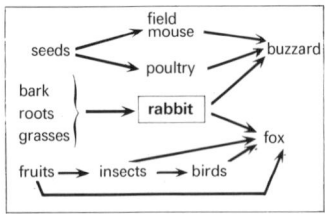

Food web around the rabbit.

Nutrient cycle.

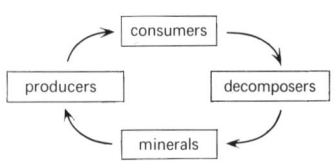

What happens when one of the links of the chain disappears, like the rabbits in the story? Not only is a food chain broken but the whole balance between prey and predators in the habitat is upset. Each living organism is usually part of several food chains forming an overall food web. Once the natural balance is destroyed, a new balance will be set up, but this may lead to the expansion of species which are considered more harmful than those destroyed (fruit-tree pests in this example). We must, therefore, be extremely careful before we decide that a species is 'harmful' and should be destroyed. Perhaps biologists should not even use the word.

Taking a more general view, we see that the first link in every food chain is always a green plant. These are the only organisms that can live by extracting from their environment just water, carbon dioxide and minerals. Green plants can make their own organic material from these ingredients, using the energy from sunlight which is captured by their chlorophyll. Thus, they are the makers of food for all living things and so we call them producers. Herbivorous animals are called primary consumers and carnivores are secondary, tertiary or fourth order consumers, according to their position in the chain. When animals or plants die, their bodies are decomposed by small animals, fungi or bacteria (decomposers) into minerals. These can then be used again by green plants, completing the nutrient cycle.

Food chains and webs show qualitatively the interdependent relationships between organisms in a habitat, but it is just as important to consider the quantitative aspect of these relationships. 90 per cent of the food eaten by every living individual is used to provide energy. This means there is a considerable loss of mass between each link in a food chain. The low rate (10 per cent on average) of conversion of food into new living matter is expressed as a food pyramid. This pyramid explains why, in any habitat, the number and especially the mass of carnivorous predators are always much smaller than the number and mass of herbivores, which in turn are much smaller than those of plants. We can also see that, because humans are placed at the top of food pyramids, nature cannot provide us with an inexhaustible supply of food unless we respect the delicate equilibria through which it is regulated. Are we not led into the thoughtless exploitation of nature's resources in order to meet the needs of an ever expanding population? Too often natural resources and energy supplies are wasted, forests destroyed, seas plundered, animal and plant species forced into extinction, nature polluted by pesticides and the wastes of our civilization. If we continue to behave in this way, will we not eventually endanger the survival of our own species? It is essential that we are all aware of these dangers, and take part in local, national and international efforts to give proper consideration to the natural resources, on which our lives depend.

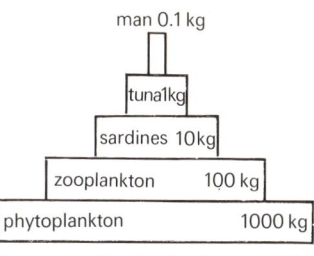

An example of a food pyramid in the sea. 1000 kg of phytoplankton are needed to produce 100 kg of zooplankton which produce 10 kg of sardines which, in turn produce 1 kg of tuna. Therefore 1 tonne (1000 kg) of phytoplankton is needed to provide 100 g of useful food for man.

The horrifying effects of pollution: crabs and starfish killed by an oil slick.

Acknowledgements

The publisher would like to thank the following for permission to reproduce their photographs:
Cover : SUNDANLE (Jacana); *page 3 :* ROBERT (Jacana); *pages 5 and 17 :* R. M. NOAILLES (Jacana); *page 6 :* ARMELLE KERNEIS (Jacana); *page 8 :* J. L. S. DUBOIS (Jacana); *pages 10 and 11 :* A. VISAGE (Jacana); *page 13 :* M. C. NOAILLES; *pages 14 and 19 :* HERVE CHAUMETON (Jacana); *page 15 :* ANDRE BERT (Jacana); *page 16 :* Jacana; *page 18 :* K. ROSS (Jacana); *page 20 :* MARCEL FRAAS (Atlas-Photo); *page 22 :* BAILLEAU (Jacana); *page 23 :* BOS (Jacana); *page 24 :* C. LENARS (Atlas-Photo); *page 25 :* CLAUDE NARDIN (Jacana); *page 26 :* R. FENAUX (Jacana); *page 27 :* KERNEIS (Jacana); *page 31 :* BERTOT (Atlas-Photo).

Translated into English by Michael Spincer.

LONGMAN GROUP LIMITED
London
Associated companies, branches and representatives throughout the world

© Librairie Vuibert, Paris, 1976
English translation © Longman Group Limited, 1980

First published in French 1976
English translation first published 1980
ISBN 0 582 32299 5

Printed in Hong Kong by
Wilture Enterprises (International) Ltd